EXPLORING OUR SOLAR SYSTEM

HOT PLANETS

MERCURY AND VENUS

DAVID JEFFERIS

Crabtree Publishing Company

www.crabtreebooks.com

■THE HOT PLANETS

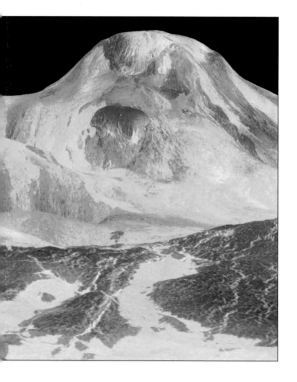

Mercury and Venus are the "hot planets" of the **solar system**. They are also the two planets that are closest to the sun. Neither of them are destinations where you would want to take a vacation. They are places of searing heat that could fry you to a crisp shortly after landing. One of the few things that Mercury and Venus have in common with our own planet, Earth, is that they have a solid, rocky surface, or **crust**. The surface of Mercury is also pitted with thousands of enormous craters, while Venus has more **volcanoes** than any other planet.

Crabtree Publishing Company

PMB 59051,
350 Fifth Avenue, 59th Floor
New York, NY 10118

616 Welland Avenue,
St. Catharines, Ontario
L2M 5V6

Published by Crabtree
Publishing Company
© 2009

Written and produced by:
    David Jefferis/Buzz Books
Educational advisor:
    Julie Stapleton
Science advisor:
    Mat Irvine FBIS
Editor: Ellen Rodger
Copy editor:
    Adrianna Morganelli
Proofreader: Crystal Sikkens
Project editor: Robert Walker
Production coordinator:
    Katherine Berti

Printed in the U.S.A./112010/AL20101101

Library and Archives Canada Cataloguing in Publication

Jefferis, David
    Hot planets : Mercury and Venus / David Jefferis.

(Exploring our solar system)
Includes index.
ISBN 978-0-7787-3735-3 (bound).--ISBN 978-0-7787-3751-3 (pbk.)

    1. Mercury (Planet)--Juvenile literature. 2. Venus (Planet)--Juvenile literature. I. Title. II. Series: Exploring our solar system (St. Catharines, Ont.)

QB611.J43 2008   j523.41   C2008-902946-1

■ **ACKNOWLEDGEMENTS**
We wish to thank all those people who have helped to create this publication. Information and images were supplied by:

Agencies and organizations:
    Adler Planetarium
    APL Applied Physics Laboratory
    CIS Carnegie Institute for Science
    ESA European Space Agency
    JHU John Hopkins University
    JPL Jet Propulsion Laboratory
    NASA National Aeronautics and Space Administration
    NSSDC National Space Science Data Center

Collections and individuals:
    Alpha Archive
    Calvin J. Hamilton
    Mat Irvine for Mariner 10 and Venera
        miniatures, evening sky photograph
    Gavin Page/Design Shop

Cosmic collision visualization inspired by the works of Fahed Sulehria.

Library of Congress Cataloging-in-Publication Data

Jefferis, David.
    Hot planets : Mercury and Venus / David Jefferis.
        p. cm. -- (Exploring our solar system)
    Includes index.
    ISBN-13: 978-0-7787-3751-3 (pbk. : alk. paper)
    ISBN-10: 0-7787-3751-9 (pbk. : alk. paper)
    ISBN-13: 978-0-7787-3735-3 (reinforced library binding : alk. paper)
    ISBN-10: 0-7787-3735-7 (reinforced library binding : alk. paper)
    1. Mercury (Planet)--Juvenile literature. 2. Venus (Planet)--Juvenile literature. I. Title. II. Series.

QB611.J44 2008
523.41--dc22

                                                                                        2008019652

CONTENTS

■ WHAT ARE HOT PLANETS?

The hot planets are the two planets closest to the sun—crater-scarred Mercury, and "hell-planet" Venus.

■ ARE THEY BIG PLANETS?

Mercury is the smallest of the eight planets. As you can see, it is not much larger than our own moon. Venus is sometimes called Earth's "evil twin". It is nearly as big as Earth, but it has a hot, poisonous **atmosphere**.

Moon

Venus

Earth

Mercury

■ Here Mercury and Venus are shown to scale with Earth and the moon. Mercury and the moon look similar, as both have surfaces covered with craters punched out by meteors, **or rocks from space.**

□ HOW DID THEY GET THEIR NAMES?

Mercury was named after the fast-moving Roman messenger of the gods. It is a name that suits it. Mercury whirls around the sun once every 88 days. Venus was named for the Roman goddess of love.

WOW!
The thick clouds of Venus reflect the sun's light well. They make Venus the brightest object in our night skies, apart from Earth's nearby **satellite**, the moon.

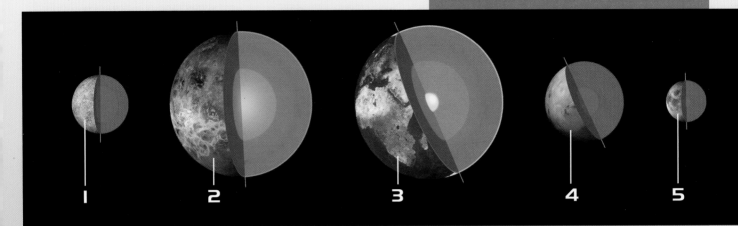

1   2   3   4   5

WHAT ARE ROCKY PLANETS?

These are the four smaller planets that **orbit** closest to the sun. They all have solid surfaces made of rock, unlike the huge planets orbiting further from the sun. Jupiter, Saturn, Uranus and Neptune are large balls of gas, with no solid surfaces on which you could walk.

■ **The insides of the** rocky planets **are similar, with a very hot central** core, **surrounded by a cooler** mantle. **The surface or crust is made of solid rock. The rocky planets are: Mercury (1), Venus (2), Earth (3), Mars (4). The moon (5) also has a rocky surface, but it is Earth's satellite, not a planet.**

HOW FAR AWAY FROM THE SUN ARE THEY?

Mercury is the 'first rock' from the sun. Its average distance is just 36 million miles (58 million km), compared with Earth, the "third rock" from the sun's orbit of 93 million miles (150 million km). Venus circles the sun about midway between the two, at 67 million miles (108 million km). The heavy atmosphere of Venus locks the heat in like a thick blanket.

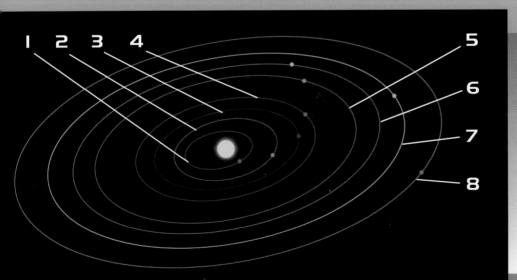

1   2   3   4   5

6

7

8

■ **Mercury and Venus are the innermost of the rocky planets. Like the other planets, they follow a curving path around the sun, called an orbit.**

1	Mercury	5	Jupiter
2	Venus	6	Saturn
3	Earth	7	Uranus
4	Mars	8	Neptune

HOW OLD ARE MERCURY AND VENUS?

Mercury and Venus are thought to have formed about 4.6 billion years ago, at about the same time as the other planets and the sun.

■ The early solar system may have looked something like this. The sun slowly heats up as it draws in more material, while young planets form from the swirling dust cloud.

■ HOW DID THEY FORM?

Scientists think that the solar system—the sun, the planets, and other space material—formed from a huge, slowly turning cloud of gas, dust, and rocks. Such star and planet formation is a process that we know goes on. Astronomers can see the process happening today in distant gas clouds in space.

■ DO MERCURY AND VENUS HAVE MOONS?

No they do not, and they are the only planets in the solar system that have none. Earth has one moon, and Mars has two small moons.

■ WHICH IS THE CLOSEST PLANET TO EARTH?

Venus is the closest. Every 584 days, Venus and Earth pass within 25 million miles (40 million km) of each other.

WOW!
In 1974, scientists thought they had found a moon of Mercury. It was briefly named "Charley", after a pet dog! But careful study showed that in reality, it was a distant star.

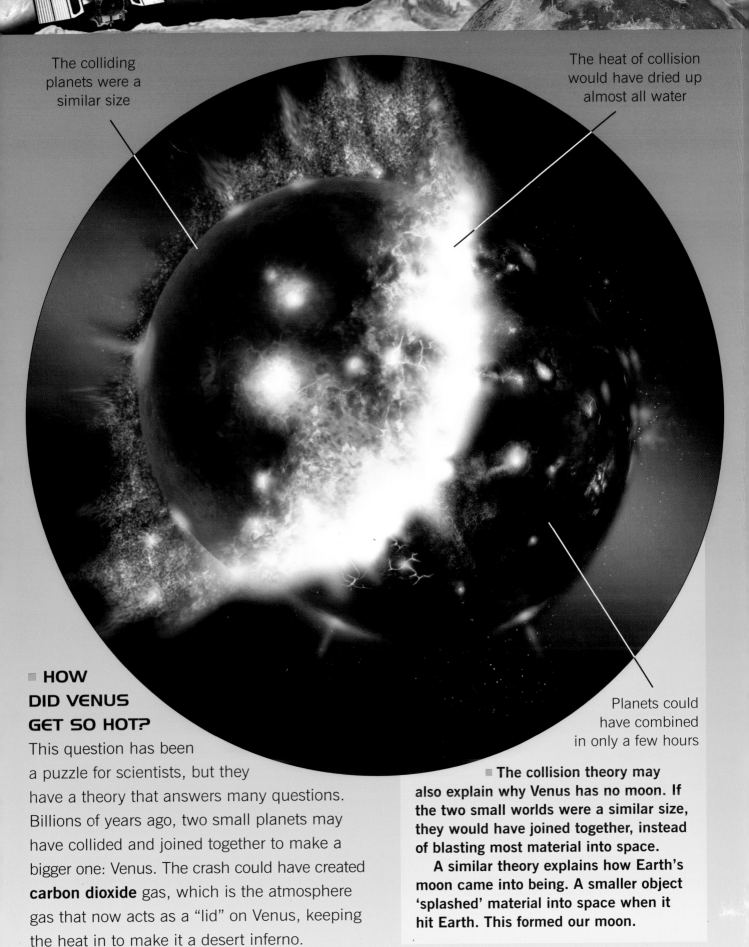

The colliding planets were a similar size

The heat of collision would have dried up almost all water

Planets could have combined in only a few hours

■ HOW DID VENUS GET SO HOT?

This question has been a puzzle for scientists, but they have a theory that answers many questions. Billions of years ago, two small planets may have collided and joined together to make a bigger one: Venus. The crash could have created **carbon dioxide** gas, which is the atmosphere gas that now acts as a "lid" on Venus, keeping the heat in to make it a desert inferno.

■ The collision theory may also explain why Venus has no moon. If the two small worlds were a similar size, they would have joined together, instead of blasting most material into space.

A similar theory explains how Earth's moon came into being. A smaller object 'splashed' material into space when it hit Earth. This formed our moon.

■HOW DO THE TWO PLANETS COMPARE?

Mercury and Venus are very different planets, although both their surfaces are covered with thousands of craters.

■ WHAT CAUSED ALL THE CRATERS?

Mercury's craters were almost all formed when the planet was struck in the distant past by meteors. Meteors were created from material that formed the sun and the planets. Most of the craters on Venus are **caldera**, gaping holes formed by the collapse of volcano mouths.

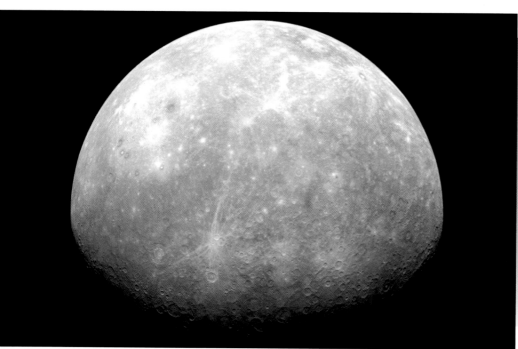

■ Countless craters mark the surface of Mercury. One detailed survey noted 763 of them across an area that measured just 172 miles (276 km) wide.
  Most of Mercury's craters are named after famous writers and artists. Craters bigger than 155 miles (250 km) across, are called basins.

■ COULD I USE A COMPASS ON MERCURY?

Yes! Like Earth, Mercury has a molten iron core that creates electric currents. These generate a magnetic field that is about 100 times weaker than Earth's. Venus has almost no magnetic field at all, perhaps because it rotates very slowly and a moving core is needed to create such a field.

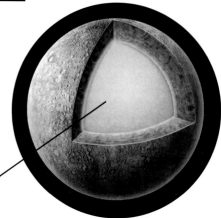

Mercury has a large iron core

WOW!
Compared to Mercury, Venus has very few small impact craters. This is because small meteors burn up in the thick atmosphere before they hit the surface.

■ These craters are in a part of Venus nicknamed the 'crater farm.' If you could see them through the thick atmosphere, they would be much paler colors. These are computer enhanced to illustrate features clearly.

■ HOW LONG IS A DAY ON THESE PLANETS?

Both planets take a very long time to rotate around their **axes**. Compared with Earth's 24-hour-long days, Mercury takes 58 days, 15.5 hours to rotate once. Venus turns even more slowly, rotating on its axis only every 243 Earth days.

□ Earth 14.7 lb/sq in (1 bar)

□ Venus 1.35 tons/sq in (92 bar)

■ At ground level, the thick atmosphere of Venus presses down at a pressure 92 times that on Earth. That is about the same as being more than half a mile (1 km) under the sea.
On Venus, you would be burned alive and squashed flat, both at the same time!

■ COULD I BREATHE THERE?

Neither world has a breathable atmosphere. Mercury has almost none at all, just a few traces of gases such as oxygen and sodium. Venus is the opposite, with a super-thick covering of mostly carbon dioxide, with traces of nitrogen. Poisonous gases are also belched out by thousands of volcanoes.

■ WHAT WAS THE FIRST PROBE TO MERCURY?

Our first close look at Mercury came in 1974, when a U.S. space probe flew past it after it flew by Venus.

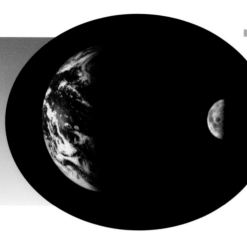

■ Early in the flight to distant Mercury, Mariner 10 took this picture after passing Earth and the moon.

■ WHY WAS A TWO-PLANET MISSION CHOSEN?

Scientists aimed Mariner 10 at Venus because it was the first spacecraft to try out a "gravity slingshot" method. This method used the **gravity** of another planet—Venus—to bend its course toward another planet—Mercury.

■ HOW LONG DID MARINER 10'S MISSION TAKE?

In all, Mariner 10's survey of the planet lasted over two years. Traveling between planets is not a quick, straight-line course, because this would use too much fuel. A slower gravity slingshot saves fuel, as an engine burn is not needed to change course.

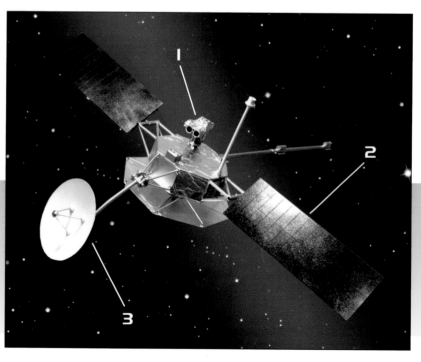

WOW!
Mariner 10's course was based on work by an Italian astronomer, Guiseppe Colombo. A new European Mercury probe is named after him.

■ The Mariner probe carried cameras and other instruments in the main body section (1). Electricity came from a pair of solar-panel 'wings' (2). A powerful radio antenna (3) sent information back to Earth.

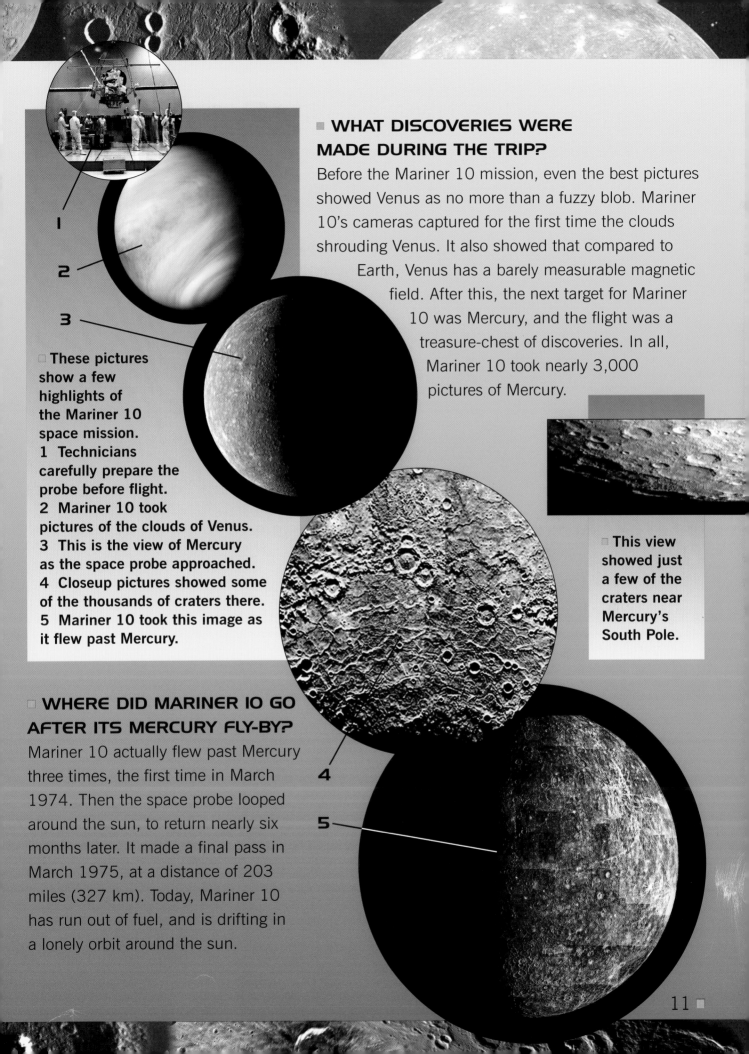

■ WHAT DISCOVERIES WERE MADE DURING THE TRIP?

Before the Mariner 10 mission, even the best pictures showed Venus as no more than a fuzzy blob. Mariner 10's cameras captured for the first time the clouds shrouding Venus. It also showed that compared to Earth, Venus has a barely measurable magnetic field. After this, the next target for Mariner 10 was Mercury, and the flight was a treasure-chest of discoveries. In all, Mariner 10 took nearly 3,000 pictures of Mercury.

■ These pictures show a few highlights of the Mariner 10 space mission.
1 Technicians carefully prepare the probe before flight.
2 Mariner 10 took pictures of the clouds of Venus.
3 This is the view of Mercury as the space probe approached.
4 Closeup pictures showed some of the thousands of craters there.
5 Mariner 10 took this image as it flew past Mercury.

■ This view showed just a few of the craters near Mercury's South Pole.

□ WHERE DID MARINER 10 GO AFTER ITS MERCURY FLY-BY?

Mariner 10 actually flew past Mercury three times, the first time in March 1974. Then the space probe looped around the sun, to return nearly six months later. It made a final pass in March 1975, at a distance of 203 miles (327 km). Today, Mariner 10 has run out of fuel, and is drifting in a lonely orbit around the sun.

1

2

3

4

5

ARE THERE SPIDERS ON MERCURY?

Mercury is a lifeless world, but the Messenger space probe of 2008 spotted an unusual feature that scientists called the "spider".

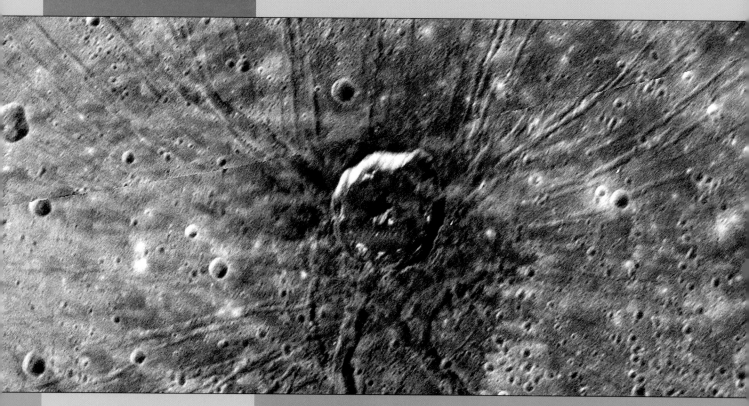

■ The spider lies on the floor of the Caloris basin, and is younger than the meteor impact that formed the basin itself. Even so, the spider is likely to be more than a billion years old.

□ WHY IS IT CALLED THE SPIDER?

Scientists thought it looked a bit like a spider with more than eight 'legs.' In fact, the legs are more than 50 troughs in the surface. The spider's 'body' is a crater, about 40 km (25 miles) across.

□ WHAT CAUSED THE SPIDER'S LEGS?

Some may have existed before, but it is likely that the meteor that made the crater made most of them.

WOW!
Mercury rotates very slowly, just once every 58.5 Earth days. It rotates exactly three times for every two complete orbits it makes around the sun.

■ These 36 pictures show how a space probe can send detailed closeups back to Earth.

A single image cannot show enough detail of the entire planet, so a number of closeups are taken instead. These are joined together to make one big image, called a photomosaic. This combined image shows far more detail than would be possible any other way.

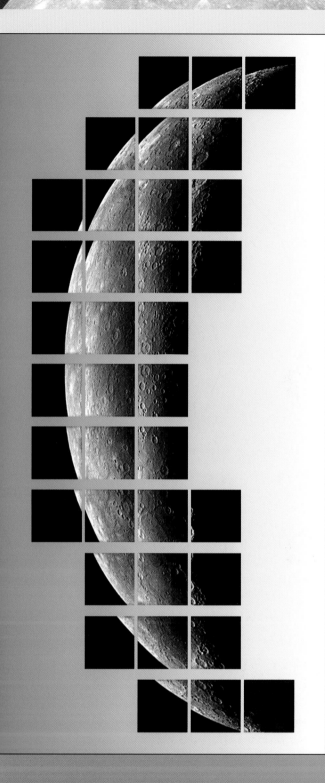

■ HOW DID WE GET THESE PICTURES?

The pictures on these pages were all taken in 2008, by the U.S. Messenger space probe. This spacecraft was the first visitor to Mercury since Mariner 10's last fly-by in 1975.

Messenger space probe

Dark halos

■ ARE THERE OTHER STRANGE FEATURES?

There are some puzzles. Near the South Pole, there are craters that are surrounded by "dark halos." This could be because the surrounding area was partly melted by the heat of the explosion when the meteor struck. Rocks that are melted suddenly in this way are often darker when they cool down afterwards.

COULD I STAND ON MERCURY'S SURFACE?

If you landed on Mercury, you might think it was Earth's moon—a barren landscape, strewn with rocks and craters, set against an airless black sky.

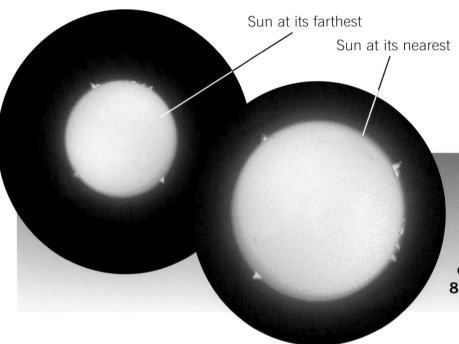

Sun at its farthest

Sun at its nearest

☐ Mercury has an odd orbit, one that is more of an oval shape than the near-circular orbit of Earth, so the sun appears to vary in size as its distance changes through the 88-day Mercurian year.

■ WHAT DOES THE SUN LOOK LIKE IN THE SKY?

The first thing you would notice is that the sun looks huge—about three times bigger than from Earth. The sun is so close that more than six times more heat and energy floods down onto Mercury's sunlit areas.

■ SO HOW HOT DOES IT GET ON THE GROUND?

It depends where you are! At noon, temperatures on Mercury can climb to 800°F (427°C), but at night, they fall hundreds of degrees below zero. Some places near Mercury's poles stay in shadow all the time. Here, there may be some water, frozen as ice in permanently shadowed parts of craters.

■ Partly hidden by a rock outcrop, the sun pours energy down on Mercury. Craters and chunks of rock cover the ground, while a few stars are bright enough to outshine the sun's glare. Two of these "stars" are actually planets—Venus is one of them, Earth the other, with the moon next to it.

■ HOW MUCH WOULD I WEIGH ON MERCURY?

Mercury is the smallest rocky world, and its gravity—the force pulling you to its surface—is much less than on Earth or Venus. If you weighed around 100 lb (45 kg) on Earth, you would be a featherweight 38 lb (17.2 kg) on Mercury. The light gravity also means that like Earth's moon, Mercury has lost almost all of its atmosphere into space.

WOW!
Mercury is the smallest rocky planet, and is also smaller than some moons. Ganymede (moon of Jupiter) and Titan (moon of Saturn) are both bigger.

■ Mercury has the biggest temperature range of all the planets. Venus has very little, because its thick atmosphere spreads heat throughout the planet.

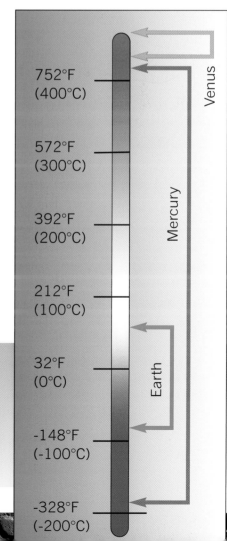

752°F (400°C)	Venus
572°F (300°C)	
392°F (200°C)	Mercury
212°F (100°C)	
32°F (0°C)	Earth
-148°F (-100°C)	
-328°F (-200°C)	

WHAT IS MERCURY'S CALORIS BASIN?

Also known as Caloris Planitia, the basin is actually an enormous meteor impact crater, one of the biggest in the solar system.

HOW BIG IS THE CALORIS BASIN?

It is huge, like a giant 'bullseye' that is 960 miles (1,550 km) across. It is surrounded by a circle of mountains 1.2 miles (2 km) high. The ground inside is mostly flat and is partly a **lava** plain, much like the surface of Earth's moon.

As the solar system formed, there were many huge chunks of rock drifting through space. Here is an idea of how things may have looked, as a meteor nears Mercury (arrowed).

WHEN DID THE IMPACT HAPPEN?

Caloris basin is very old. Space scientists believe it was formed about 3.5 billion years ago. Its bullseye appearance was caused by material blasted out by the meteor bouncing back down on to the surface again after the first impact.

Mercury has thousands of craters formed by meteor impacts. These pictures show how they were made.
1 Meteor falls at high speed.
2 Meteor smashes into the surface, sending shock waves into the ground.
3 The explosion blows out a hole, with material thrown all around.
4 A basin surrounded by a circular rim wall is what remains.

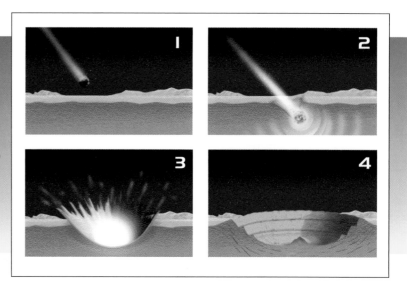

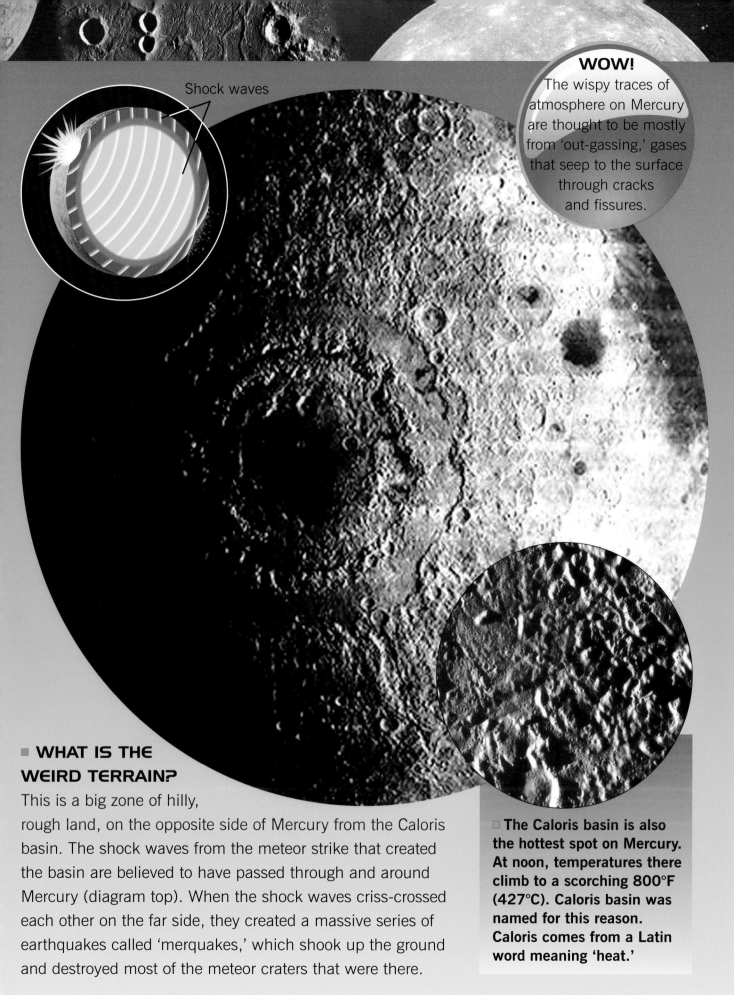

Shock waves

WOW!
The wispy traces of atmosphere on Mercury are thought to be mostly from 'out-gassing,' gases that seep to the surface through cracks and fissures.

■ WHAT IS THE WEIRD TERRAIN?

This is a big zone of hilly, rough land, on the opposite side of Mercury from the Caloris basin. The shock waves from the meteor strike that created the basin are believed to have passed through and around Mercury (diagram top). When the shock waves criss-crossed each other on the far side, they created a massive series of earthquakes called 'merquakes,' which shook up the ground and destroyed most of the meteor craters that were there.

☐ **The Caloris basin is also the hottest spot on Mercury. At noon, temperatures there climb to a scorching 800°F (427°C). Caloris basin was named for this reason. Caloris comes from a Latin word meaning 'heat.'**

■ WHY IS VENUS CALLED THE 'HELL PLANET?'

Conditions on Venus are what many believe hell might be like. It is an unbelievably hot and bone-dry planet.

Plate-like rocks   Grit and stones   Part of Venera's landing section

■ WHAT IS THE SURFACE LIKE THERE?

The surface of Venus is definitely not pleasant. Temperatures are hot enough to melt lead. The soupy atmosphere is thick enough to squash you. The clouds are droplets of burning sulphuric acid.

WOW!
Venus turns very slowly. It takes 243 Earth days to rotate around its axis. It takes 224.7 Earth days to orbit the sun, so a Venusian day is longer than its year is.

■ WHAT IS THE AIR MADE OF?

It is made mostly of carbon dioxide gas, which traps some of the sun's energy, as well as heat from the planet's volcanoes. This heat trapping, or **greenhouse effect**, has turned Venus into the hottest planet.

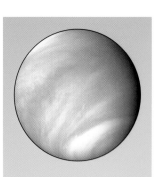

■ The thick clouds of Venus meant that we had no idea what lay under them until space probes were sent to find out.

■ The picture below was taken by the Russian Venera 13 space probe that landed in 1982.

Russia landed nine space probes on Venus, but none of them lasted long. Even the successful Venera 13 (right), survived just 127 minutes before being crushed by the atmosphere.

■ HAVE SPACE PROBES LANDED?

Russia has sent several Venera space probes to land on the surface. The first success was Venera 7, which landed in December 1970. It was the first space probe to make a safe landing on another planet, and the first to send information to Earth from the surface.

■ HOW LONG DID THEY LAST?

Conditions on Venus are so extreme that none of Russia's landers lasted very long. Venera 7 managed to send weak signals for only 23 minutes before the equipment failed.

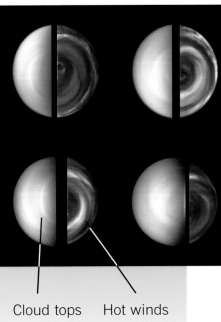

■ These pictures combine images of the cloud tops (blue) with the swirl of winds below (red). Temperatures on Venus reach over 840°F (450°C). Nights are no cooler, because the high winds carry heat all around Venus.

Cloud tops   Hot winds

ARE THERE HURRICANES ON VENUS?

There certainly are! The atmosphere has winds that blow at 225 mph (360 km/h) or more.

■ **This view of the North Pole of Venus shows a spinning double-hurricane.**
Hurricanes on Earth are single weather systems. On Venus, the swirling "polar vortex" **has two, with winds blowing in an S-shape.**

■ WHAT IS SUPER-ROTATION?

This is the name for the high-speed winds that blow the entire atmosphere around the planet in just four days or so. The super-rotating winds blow heat from dayside to nightside, so the long nights of Venus are just as blistering hot as its days are.

■ DO GALES BLOW EVERYWHERE ON VENUS?

Surprisingly, not on the surface, where the massive weight of the atmosphere slows things down dramatically. Down at ground level, the atmosphere is so thick that winds there are very sluggish, blowing at just 8 mph (12 km/h).

■ **These rocks have sharp edges because surface winds are too gentle to wear them down.**

■ Space probes have been the key to studying Venus. This probe is the European Venus Express, which has provided a lot of information, including details of the polar vortex. As the picture sequence above shows, the shape of each vortex changes all the time in ways that are not yet fully understood.

■ IS THERE ANYWHERE ON VENUS THAT IS NOT QUITE SO HARSH?

Not on the surface or anywhere in the lower air layers. Things do change higher up, where the atmosphere starts to get thinner, just as it does on Earth. In fact, at heights of about 30-40 miles (50-65 km), pressures and temperatures are similar to Earth's at ground level. It is still not a breathable gas mixture. Despite this, this part of Venus's atmosphere is the most Earth-like environment in the solar system.

HOW MANY VOLCANOES ARE ON VENUS?

Venus has more volcanoes than any planet in the solar system. Many of them are huge, at more than 100 miles (162 km) across.

ARE VOLCANOES ERUPTING ON VENUS AT PRESENT?

Most of the 1,600-plus volcanoes on Venus are thought to be extinct. Maat Mons (left) may still be active. It is the tallest volcano on Venus, and pictures show signs of recent ash flows at the summit and down the slopes.

◻ Maat Mons has a big crater, or caldera, at its top. This is 19 miles (31 km) across, and there are five smaller craters nestling inside.

◻ If you could visit Venus, you might see a view like this. Here Maat Mons belches out steam, smoke, ash, and hot rivers of lava.

WHAT DO VENUSIAN VOLCANOES LOOK LIKE?

Most of them are **shield volcanoes**, named for their wide, gently rising shape. They are formed by molten lava spreading out and cooling as it goes to form an ever-larger zone of solid rock. Eventually, the lava builds up into a wide, low mountain.

◻ Russian Venera space probes recorded the sounds of Venusian thunder with microphones.

WOW!
Venus has lots of volcanoes, but they are not the solar system's biggest. Olympus Mons on Mars is three times higher than anything on Venus.

ARE VOLCANOES ON VENUS LIKE THOSE ON EARTH?

There are shield volcanoes on Earth, but not as big. The highest is Mauna Kea, which rises 6.06 miles (9.75 km) from the floor of the Pacific Ocean.

■ HOW BIG ARE THE VOLCANOES?

Many of the shield volcanoes on Venus spread out across wide areas, but few of them are very tall. Most rise less than 1 mile (1.6 km) above the surrounding plains. The biggest volcano, Maat Mons, rises higher than normal. It has a summit about 5 miles (8 km) high.

■ We know that there are bursts of lightning on Venus, but we are not quite sure whether they also appear at ground level. Some space-probe results suggest that most lightning may be inside clouds about 22 miles (35 km) up.

WHAT ARE ARACHNOIDS?

Arachnoids are markings on the surface of Venus. They are called arachnoids because they look a bit like a spider's web.

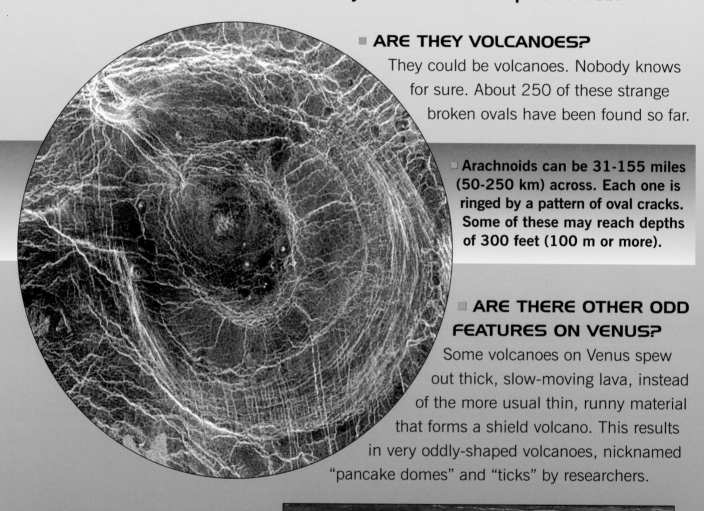

ARE THEY VOLCANOES?

They could be volcanoes. Nobody knows for sure. About 250 of these strange broken ovals have been found so far.

Arachnoids can be 31-155 miles (50-250 km) across. Each one is ringed by a pattern of oval cracks. Some of these may reach depths of 300 feet (100 m or more).

ARE THERE OTHER ODD FEATURES ON VENUS?

Some volcanoes on Venus spew out thick, slow-moving lava, instead of the more usual thin, runny material that forms a shield volcano. This results in very oddly-shaped volcanoes, nicknamed "pancake domes" and "ticks" by researchers.

Pancake domes (right) rise less than a mile or so above the surrounding land. They occur in groups. The many 'legs' of ticks (right) are ridges and valleys, caused by hot, flowing lava and landslides.

HOW WAS THIS GLOBE OF VENUS MADE?

It is a photomosaic of shots taken by the Magellan probe. It had an instrument that used radio waves to penetrate the thick clouds.

COULD THERE BE LIFE ON VENUS?

Almost certainly not, but on Earth, creatures called **extremophiles** manage to survive in hot geysers, which are deadly to other forms of life. It is possible that micro-organisms may live in the upper-air layers of Venus, where conditions are not as severe as on the surface.

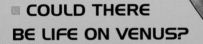

■ These are not the true colors of Venus. Instead they are computer-coded to show high and low areas. Blues and greens show the lowest land, brown and whites are hilly areas.

□ Magellan was a U.S. space probe that mapped most of the surface of Venus. It stayed in orbit to study the planet from 1990 to 1994.

WOW!
Only Venera probes have landed on Venus. Conditions there are so tough that later probes, such as Magellan and Venus Express, have studied the planet from space.

■ CAN I OBSERVE THE HOT PLANETS?

Venus is often the brightest object in the night sky, apart from the moon. Mercury is more difficult to see, because it is so close to the sun.

■ WHAT IS A GOOD WAY TO SPOT PLANETS?

Astronomers require telescopes to help them view distant planets, but Venus and Mercury are too far away to see any details of their surfaces. Amateur telescopes cannot show as much as big observatory instruments can.

DANGER
NEVER look at the sun directly. Doing so will cause eye damage or blindness. It is possible to project the sun's image on a card, but **ALWAYS** get an adult to help.

■ WHAT CAN I SEE?

We used a six-inch telescope (right) to take the Venus pictures above. They were taken over a period of six months. They show Venus getting bigger as it gradually approaches Earth. All the shots were then combined into one using a computer program.

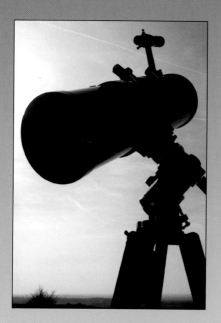

■ You will need at least a six-inch telescope to view distant space objects.
You also need a steady tripod to hold the telescope. Tripods hold the telescopes and cameras better than human hands can. Without one, the slightest tremor will make the image blurry.

■ WHAT ABOUT SPOTTING MERCURY?

It is more difficult to spot Mercury than it is to find Venus. Mercury always appears near the sun, and is easily lost in the glare. Mercury can usually be seen only around sunrise or sunset, when the sun is not so bright.

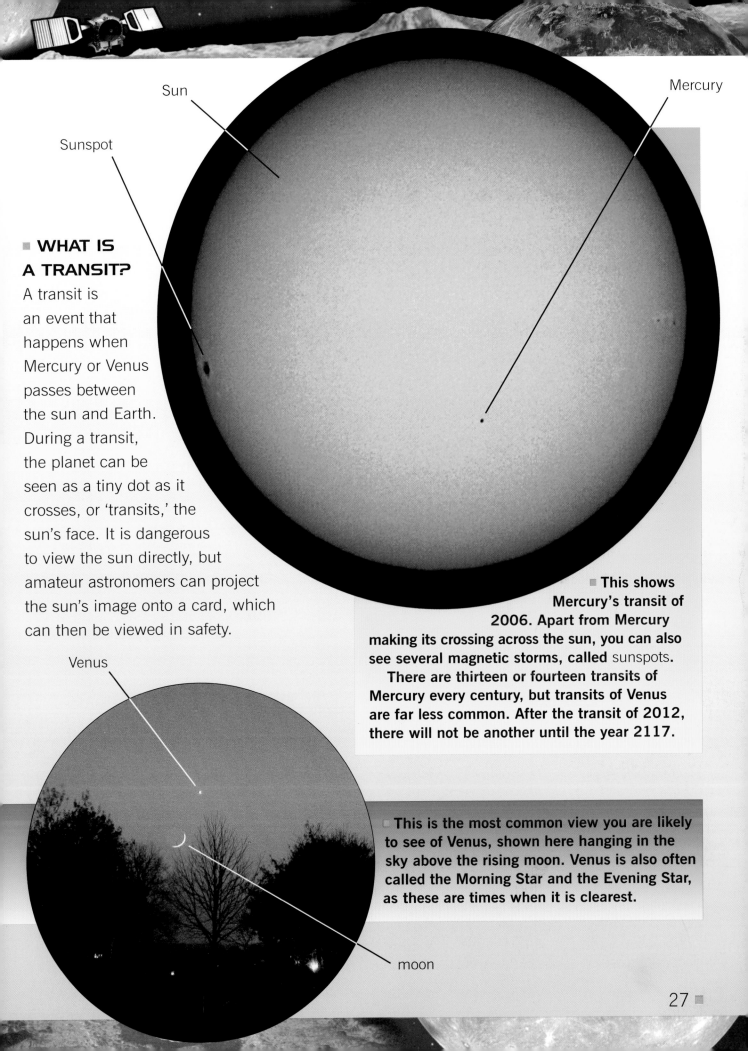

Sunspot

Sun

Mercury

■ WHAT IS A TRANSIT?

A transit is an event that happens when Mercury or Venus passes between the sun and Earth. During a transit, the planet can be seen as a tiny dot as it crosses, or 'transits,' the sun's face. It is dangerous to view the sun directly, but amateur astronomers can project the sun's image onto a card, which can then be viewed in safety.

■ **This shows Mercury's transit of 2006. Apart from Mercury making its crossing across the sun, you can also see several magnetic storms, called** sunspots. **There are thirteen or fourteen transits of Mercury every century, but transits of Venus are far less common. After the transit of 2012, there will not be another until the year 2117.**

Venus

☐ **This is the most common view you are likely to see of Venus, shown here hanging in the sky above the rising moon. Venus is also often called the Morning Star and the Evening Star, as these are times when it is clearest.**

moon

■ FACTS AND FIGURES

■ MERCURY STATISTICS

Diameter

3,030 miles (4,876 km), making Mercury just over one-third as wide as Earth.

Time to rotate (day)

Mercury rotates once on its axis every 58 Earth days and 15.5 hours.

Time to orbit once around the sun (year)

Mercury completes one orbit of the sun every 88 Earth days.

Distance to the sun

36 million miles (58 million km) average. The orbit is not circular, and at its closest, Mercury goes as close as 29 million miles (46 million km).

Composition

Mercury seems to have a metal-rich core, making up about 60 percent of the planet's **mass**. Much of the surface is covered with lava plains.

Temperature

At the Caloris basin, about 800°F (427°C). At the poles, a chilly -361°F (-183°C).

Surface gravity

Here on Earth, we live under a force of one gravity, or 1G. Mercury has a gravity pull of little more than one-third of this.

Atmosphere

Mercury has little or no atmosphere, just traces of gases seeping through cracks in the crust. These include tiny amounts of oxygen, sodium, hydrogen and helium.

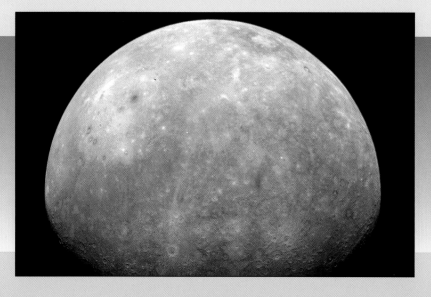

□ **Mercury looks like Earth's moon, with impact craters and large areas covered with smooth lava plains.**

The plains were made when molten material from the mantle flooded out when the crust was cracked open by meteors billions of years ago.

■ VENUS STATISTICS

Diameter

7,520 miles (12,100 km), making Venus almost as big as Earth.
Of all the planets, it is the closest to Earth's size.

Time to rotate (day)

Venus rotates very slowly, just once every 243 Earth days.

Time to orbit once around the sun (year)

Venus completes one orbit of the sun every 224.7 Earth days.

Distance to the sun

67 million miles (108 million km) average.

Composition

Venus has a metal core, similar to Mercury or Earth, though perhaps
not as large. Most of the surface is covered with plains of volcanic rock.

Temperature

The surface temperature on Venus is much the same across the planet,
about 864°F (462°C). The only cooler parts are high in the atmosphere.

Surface gravity

Slightly more than 90 percent of Earth's gravity.

Atmosphere

Extremely thick, mostly carbon dioxide with some nitrogen and clouds of
sulphur dioxide. The air pressure is 92 times more than on Earth.

□ **Venus rotates in an east-to-west direction, unlike Earth and other planets, which all turn the other way.**
Called retrograde motion, **it makes a Venusian sunrise unusual, compared with Earth. On Venus the sun seems to rise in the west, and set (right) in the east.**

GLOSSARY

Explanations for many of the terms used in this book.

Atmosphere The layers of air surrounding a planet such as Venus.
Axis Imaginary line between the poles of a planet around which it rotates.

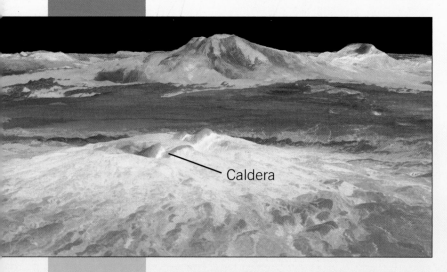

— Caldera

■ **Most of the craters on Venus are volcanic calderas, rather than ones made by meteor impacts, as they are on Mercury.**

Caldera A large volcanic crater, usually formed when the top of a volcano collapses after an eruption.
Carbon dioxide A colorless gas that makes up most of the Venusian atmosphere. On Earth, it is given off as waste in animal breath and absorbed by plants.
Core The center of a planet. The cores of Venus and Mercury are thought to be made mostly of iron.
Crust The outermost layer of rocky planets such as Mercury and Venus.

Extremophile A living thing that can survive in places deadly to other forms of life, such as humans. They are microscopic organisms.
Gravity The universal force of attraction between all objects.
Greenhouse effect The trapping of heat in a planet's atmosphere, typically by carbon dioxide gas.
Impact crater A crater formed by a meteor hitting the ground.
Lava Molten rock spewed out from a volcano. Lava may flow as a fiery liquid or as a slower mass.
Mantle The part of a planet that lies between the crust and the core.
Mass The matter an object contains.

■ **Sapas Mons is a shield volcano on Venus. It is very wide, stretching some 250 miles (400 km) across, but not very tall. Its summit is just 0.9 miles (1.5 km) high.**

Part of the sun to the same scale as the planets

Here are the sun and planets:
1 Mercury
2 Venus
3 Earth
4 Mars
5 Jupiter
6 Saturn
7 Uranus
8 Neptune

Meteor A lump of rock drifting in space. If a meteor hits the ground, it usually creates an impact crater.

Orbit The curving path a space object takes around a more massive one, such as a planet orbiting the sun.

Retrograde motion The east-to-west rotation of Venus on its axis. The other planets, including Earth, rotate in the opposite direction.

Rocky planet A planet with a solid crust. Mercury, Venus, Earth, and Mars are all rocky planets.

Satellite A space object that orbits a bigger one. Satellites can be natural, such as Earth's moon, or artificial, such as a space probe studying a planet from orbit.

Shield volcano A broad, domed volcano with gently sloping sides. It is caused by a steady buildup of lava over many years.

Solar panel A flat panel that changes the energy in sunlight to electricity.

Solar system The name for the sun and the eight planets, dwarf planets, moons, and other space objects that circle it.

Volcano A mountain with a crater at its peak through which lava, gas, and dust may spurt out, rushing up from the mantle far below.

Vortex A swirling whirlpool of gas or liquid. Venus has them at each pole.

GOING FURTHER

Using the Internet is a great way to expand your knowledge of the hot planets.

Your first visit should be to the site of the U.S. space agency, NASA. Its site shows almost everything to do with space, from the history of spaceflight to astronomy, and also plans for future missions.

There are also websites that give detailed space information. Try these sites to start with:

http://www.nasa.gov — A massive site.
http://www.space.com — Lots of stuff.
http://messenger.jhuapl.edu/ — Mercury space probe.
http://http://www.esa.int — European Space Agency.
http://www.spacedaily.com — Good for space news.

■INDEX

BRAVERY SOUP

by Maryann Cocca-Leffler

Albert Whitman & Company
Chicago, Illinois

Also by Maryann Cocca-Leffler:

JUNGLE HALLOWEEN

MISSING: ONE STUFFED RABBIT

MR. TANEN'S TIES

MR. TANEN'S TIES RULE!

MR. TANEN'S TIE TROUBLE

Library of Congress Cataloging-in-Publication Data

Cocca-Leffler, Maryann, 1958-
   Bravery soup / written and illustrated by Maryann Cocca-Leffler.
   p. cm.
   Summary: Carlin, who is frightened by everything, wants to try some of
   Big Bear's Bravery Soup, but first he must travel through a dark forest to
   a monster's cave to retrieve an important ingredient.
   ISBN 0-8075-0870-5 (hardcover) ISBN 0-8075-0871-3 (paperback)
   [1. Fear—Fiction. 2. Bears—Fiction.] I. Title. PZ7.C638 Br 2002 [E]—dc21 2001004167

The paintings are rendered in acrylic.
For more information about Albert Whitman & Company,
visit our web site at www.albertwhitman.com.
For more information about Maryann Cocca-Leffler,
visit her web site at www.maryanncoccaleffler.com.

In memory of my dear friend Marcia Sommers,
who taught me about bravery.

Carlin was afraid of everything.
He was afraid of bumps in the night,

of trying new things,

of being alone.

He was afraid of his own shadow!

"You need a bit of bravery,"
said his friend Zack, "and I know
where you can get it."

Early the next morning Zack led Carlin to the edge of the woods.
There they saw Big Bear, the bravest animal in all the land.
He was standing by the fire, stirring a big pot of soup.

"So you want some bravery, do you?" asked Big Bear.

"Y-y-yes," stammered Carlin.

"Well, I'm mixing up a batch of Bravery Soup right here, but I'm missing an important ingredient. Will you get it for me?" asked Big Bear.

"Will it make me brave?" asked Carlin.

"Most certainly," said Big Bear. "But your journey will not be easy. You must go, alone, through the Forbidden Forest to Skulk Mountain. On the mountaintop you will see a cave. In the cave you will find a box. Bring that box to me."

Carlin gasped.

"ALONE? THE FOREST? THE CAVE?"

His knees were shaking.
"You are braver than you think," said Big Bear.

When the animals heard that Carlin was venturing
into the Forbidden Forest alone, they gathered around him.

"Here is a basket of
food. The forest is full of
poisonous plants."

"Here is armor to protect
you from the wild beasts."

"Here is a raft to cross
the raging river."

"Here is a big stick to fight the fierce monster that lives in the cave."

"WILD BEASTS?
POISON?
RAGING RIVER?
FIERCE MONSTER?"

Carlin's whole body shuddered.
Then he remembered Big Bear's words:
"You are braver than you think."

"Now or never," Carlin thought.
Slowly, he walked into the thick forest.

His friends waited for a while, growing more and more worried.

"It is much too dangerous for little Carlin," they said. They decided to search for him.

As Carlin's friends entered the forest, they came across the armor.

"OH! HE IS HURT!"

They found the basket of food, untouched.

"OH, MY! HE IS STARVING!"

Then they spotted the raft,
floating at the river's edge.

"OH, NO!
THE RIVER HAS SWEPT HIM AWAY!"

But Carlin was not hurt.
He had soon realized that he could not walk fast
wearing the heavy armor.
"I can flee the beasts more quickly
without it," he thought. So he took
the armor off.

Carlin was not starving.

As he walked along, he noticed animals and birds feasting on fruits in the Forbidden Forest.

He saw some strawberries. "I can eat these!" he thought. So he dropped his heavy basket and ate the delicious fruit.

Carlin had not been swept away.

When he came to the raging river, he found that a tree lay over the water. He tossed the raft aside. Carefully, he began to make his way across. He took one tiny step and then another, until finally he reached the other side.

"I DID IT!"

he cheered.

Carlin looked up.
There before him was
Skulk Mountain.

Carlin trudged on. The mountain
was covered with thick bushes and vines.
He broke his stick cutting a path.
Then he spotted giant footprints!
Carlin had discovered the
monster's cave.

Trembling, he entered.

Minutes later, his friends found the broken stick by the mouth of the cave.

"OH, NO! HE HAS BEEN EATEN BY THE MONSTER!"

But Carlin had not been eaten by the monster.
He peered into the cave and squinted his eyes. There, on the ground,
was the box! Just then, something moved from the shadows.

THE MONSTER!

Terrified, Carlin hid behind a pile
of rocks. Suddenly, one of the
rocks tumbled over!

The sound made the monster jump and howl!

"Who is it?" he whimpered. **"WHAT D-D-DO YOU WANT?"**

Carlin had cast a long, spooky shadow.

"Could this scary monster be afraid of **ME?**" he thought.

"My name is Carlin," Carlin said in his deepest, loudest voice. "I came for the box."

"This box?" The monster tossed the box towards Carlin. "Take it!" he said.

"Just please don't hurt me!"

Carlin grabbed the box and ran from the cave ... bumping right into his friends.

"RUN! THE MONSTER!"

They all raced down the mountain,

over the river,
through the forest,

and back to the edge
of the woods.

At last, the tired animals gathered around Big Bear's fire. Carlin carried the precious box.

"Now it is time," said Big Bear.
"Open the box."
Carlin pried off the lid.

THE BOX WAS EMPTY!

Carlin was sad. "I'm sorry, Big Bear. I didn't get the secret ingredient for bravery."

"But you did," said Big Bear with a smile.
"The box was always empty."

"Do you mean my journey was for **NOTHING?**"
asked Carlin in surprise.

"Your journey was not for nothing,"
said Big Bear. "You faced the forest and
you faced your fear. It is not what is inside
the **BOX** that makes bravery. It is what is
inside of **YOU!**"

"So what about the Bravery Soup?"
asked Carlin.

"It's called Bravery Soup because
I only serve it to the brave!"
said Big Bear.

"I AM BRAVE!"

said Carlin.

And he held up his bowl.